DISCOVER YOUR OWN VARIABLE STAR

by

Martin P Nicholson

OGLE-II DIA Photometry
Field: CAR_SC1 StarID: 154935

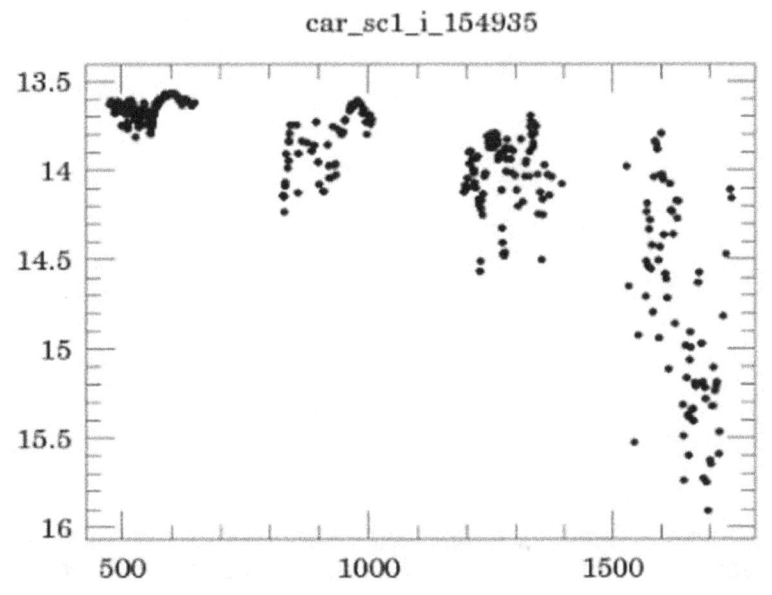

car_sc1_i_154935

Martin Nicholson
Church Stretton
Shropshire SY6 7DQ
United Kingdom

Email – newbinaries@yahoo.co.uk

The second phase of the Optical Gravitational Lensing Experiment (Udalski *et al.* 1997) was conducted from the Las Campanas Observatory in Chile using a 1.3-meter telescope operating at an effective focal ratio of *f* / 9.2. This gave an image scale of 0.417 arcsec/pixel. All observations were made in driftscan mode with the majority of observations made using an *i*-band filter with a passband near to Cousins *I*.

IDENTIFYING VARIABLE STAR TARGETS IN **OGLE**

The OGLE database can be accessed at:-

http://ogledb.astrouw.edu.pl/~ogle/photdb/phot_query.html

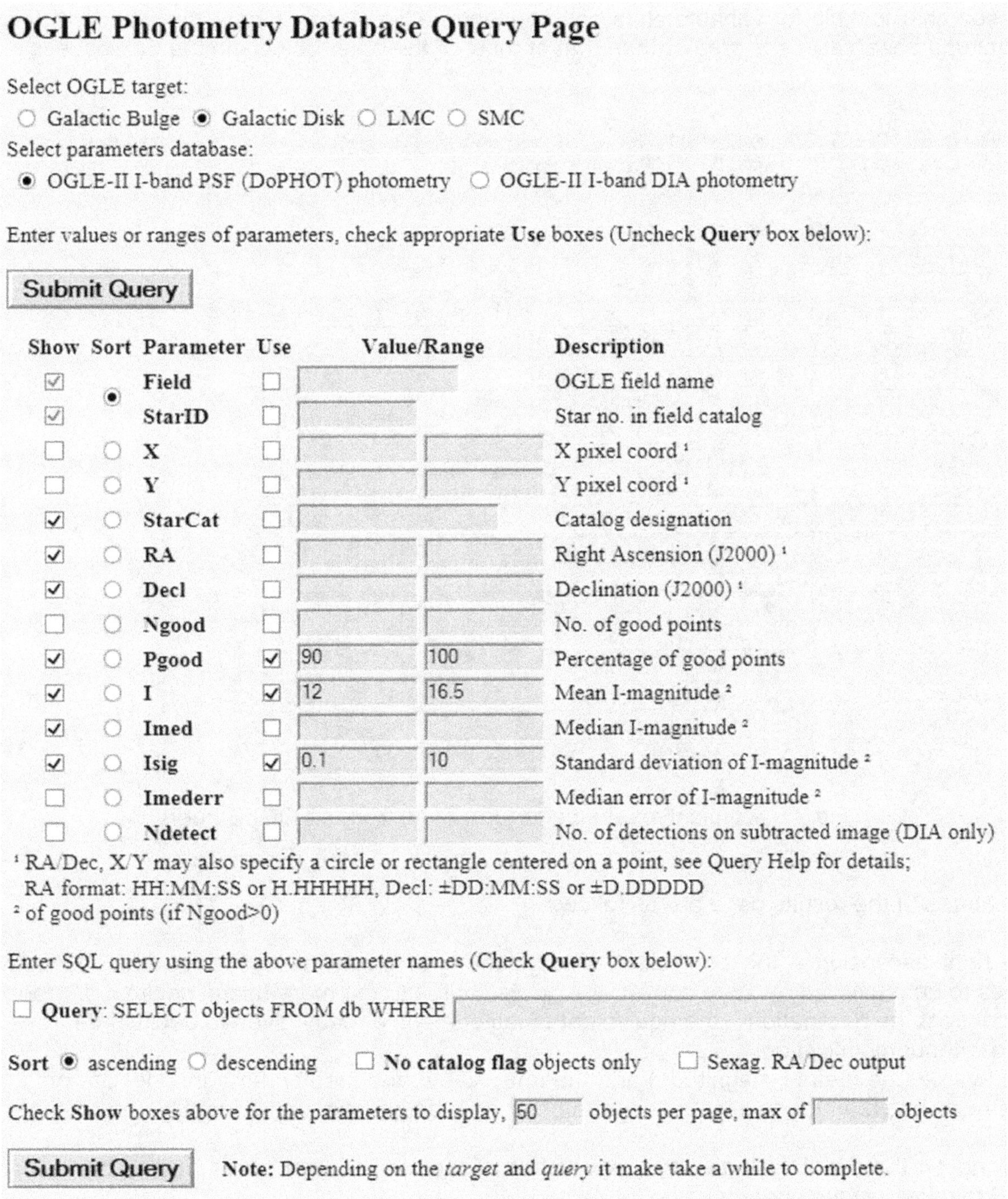

OGLE Photometry Database Query Page

Select OGLE target:
  ○ Galactic Bulge ● Galactic Disk ○ LMC ○ SMC
Select parameters database:
  ● OGLE-II I-band PSF (DoPHOT) photometry ○ OGLE-II I-band DIA photometry

Enter values or ranges of parameters, check appropriate **Use** boxes (Uncheck **Query** box below):

Submit Query

Show	Sort	Parameter	Use	Value/Range		Description
☑	●	**Field**	☐			OGLE field name
☑		**StarID**	☐			Star no. in field catalog
☐	○	**X**	☐			X pixel coord [1]
☐	○	**Y**	☐			Y pixel coord [1]
☑	○	**StarCat**	☐			Catalog designation
☑	○	**RA**	☐			Right Ascension (J2000) [1]
☑	○	**Decl**	☐			Declination (J2000) [1]
☐	○	**Ngood**	☐			No. of good points
☑	○	**Pgood**	☑	90	100	Percentage of good points
☑	○	**I**	☑	12	16.5	Mean I-magnitude [2]
☑	○	**Imed**	☐			Median I-magnitude [2]
☑	○	**Isig**	☑	0.1	10	Standard deviation of I-magnitude [2]
☐	○	**Imederr**	☐			Median error of I-magnitude [2]
☐	○	**Ndetect**	☐			No. of detections on subtracted image (DIA only)

[1] RA/Dec, X/Y may also specify a circle or rectangle centered on a point, see Query Help for details;
  RA format: HH:MM:SS or H.HHHHH, Decl: ±DD:MM:SS or ±D.DDDDD
[2] of good points (if Ngood>0)

Enter SQL query using the above parameter names (Check **Query** box below):

☐ **Query**: SELECT objects FROM db WHERE [   ]

Sort ● ascending ○ descending   ☐ **No catalog flag** objects only   ☐ Sexag. RA/Dec output

Check **Show** boxes above for the parameters to display, 50 objects per page, max of [   ] objects

Submit Query   Note: Depending on the *target* and *query* it make take a while to complete.

Fig.1 – The OGLE Photometry Database Query Page

At first sight the range of options available to the data miner appears rather overwhelming but there are a number of basic principles that simplify the task.

1. In order to achieve a useful light curve only include objects with at least 90% of observations considered reliable.
2. Faint objects, that is those nearer the detection limit of the telescope, although far more numerous, will always demonstrate more random scatter than brighter objects. This makes them harder to study.
3. A researcher looking for variable stars can concentrate on objects where the standard deviation of the magnitude observations is greater than that of most of the other objects in the database.

Phot Query:SELECT FROM disk2 WHERE Pgood>=90 and Pgood<=100 and I>=12 and I<=16.5 and Isig>=0.1 and Isig<=10
Displaying Page 1 of 194: objects 1-50 of 9685 got in 0.00 seconds

Page: Prev 1 2 3 4 5 6 7 8 9 ... Next

Download data | Options: ● Star Table or ○ Photometry ☑ good points only ● DIA or ○ PSF

No	Field	StarID	StarCat	RA	Decl	Pgood	I	Isig	Imed
1	CAR_SC1	87	110524.86-615213.8	11.090238	-61.87050	99	16.243	0.274	16.229
2	CAR_SC1	88	110520.39-615209.6	11.088996	-61.86932	97	15.984	0.116	15.964
3	CAR_SC1	89	110514.80-615208.6	11.087445	-61.86904	96	16.485	0.173	16.471
4	CAR_SC1	93	110517.24-615203.6	11.088123	-61.86766	94	16.346	0.116	16.288
5	CAR_SC1	114	110525.84-615112.6	11.090512	-61.85350	92	16.330	0.111	16.279
6	CAR_SC1	122	110516.01-615054.0	11.087782	-61.84833	97	16.068	0.149	16.061
7	CAR_SC1	131	110515.70-615038.1	11.087694	-61.84391	96	15.737	0.106	15.732
8	CAR_SC1	157	110514.24-614939.4	11.087288	-61.82762	91	15.971	0.109	15.933
9	CAR_SC1	191	110532.65-615217.4	11.092404	-61.87151	99	16.453	0.137	16.445
10	CAR_SC1	2704	110514.12-614631.5	11.087254	-61.77541	95	15.913	0.161	15.902
11	CAR_SC1	2872	110520.49-614615.7	11.089026	-61.77104	99	16.445	0.106	16.411
12	CAR_SC1	5494	110511.60-614457.1	11.086556	-61.74921	95	16.345	0.136	16.302
13	CAR_SC1	5496	110523.11-614450.4	11.089753	-61.74732	98	15.884	0.163	15.878
14	CAR_SC1	5497	110526.14-614449.1	11.090594	-61.74698	95	16.024	0.113	15.975
15	CAR_SC1	8184	110520.72-613911.3	11.089089	-61.65315	90	14.150	0.178	14.142

Fig.2 – A typical results page (part only) from an OGLE query

The main features of the results page are as follows:-

1. The right ascension – the equivalent of longitude – is given in decimal hours. The number quoted needs to be multiplied by 15 to convert the quoted value to the more useful decimal degrees.
2. By contrast the declination – the equivalent of latitude – is already given in decimal degrees and can be used without modification.
3. The I value (the mean I-magnitude) and the Imed value (the median I-magnitude) are not identical. The difference between the two values can be used as a diagnostic tool to identify certain types of variable star.
4. Clicking on the StarID will give the researcher access to the light curve and to the individual results used to construct the curve.

The next stage of the analysis is crucial and needs to be done carefully if any useful results are to be obtained.

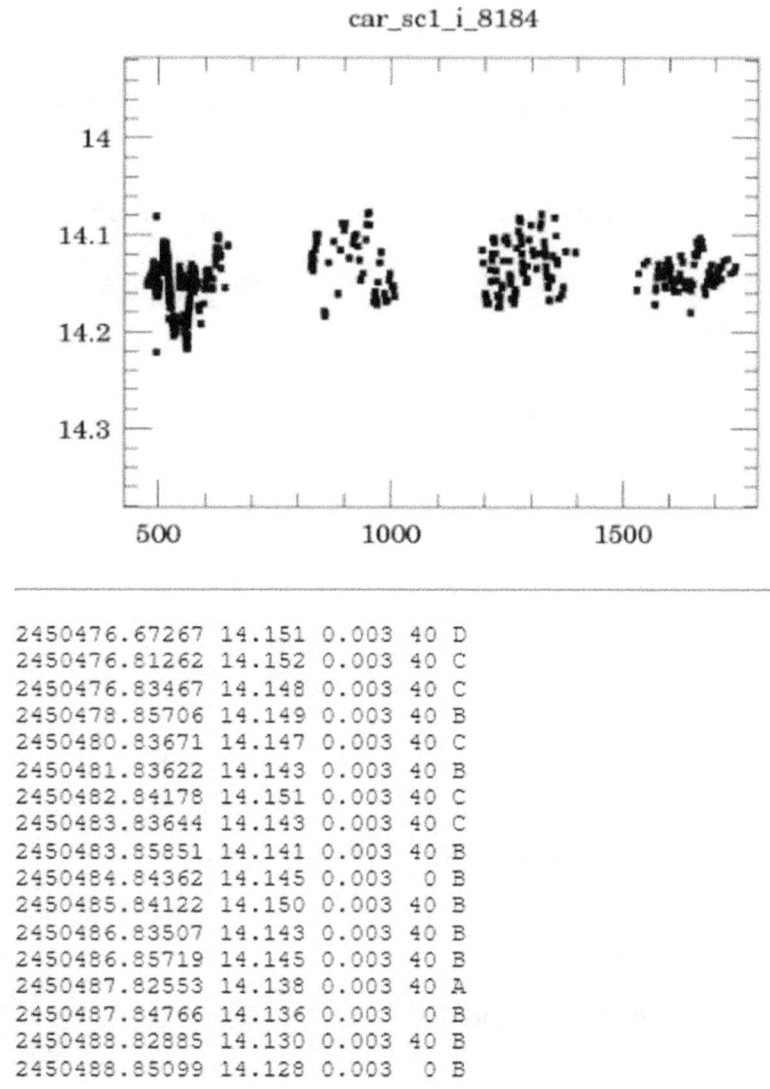

car_sc1_i_8184

2450476.67267 14.151 0.003 40 D
2450476.81262 14.152 0.003 40 C
2450476.83467 14.148 0.003 40 C
2450478.85706 14.149 0.003 40 B
2450480.83671 14.147 0.003 40 C
2450481.83622 14.143 0.003 40 B
2450482.84178 14.151 0.003 40 C
2450483.83644 14.143 0.003 40 C
2450483.85851 14.141 0.003 40 B
2450484.84362 14.145 0.003   0 B
2450485.84122 14.150 0.003 40 B
2450486.83507 14.143 0.003 40 B
2450486.85719 14.145 0.003 40 B
2450487.82553 14.138 0.003 40 A
2450487.84766 14.136 0.003   0 B
2450488.82885 14.130 0.003 40 B
2450488.85099 14.128 0.003   0 B

Fig.3 – A typical light curve showing periodic the changes in variability for star #8184

Beneath the graph are find 5 columns of data. The results need to be pasted into a spreadsheet in order to process them.

The first column is the date of the observation. Make certain that the date is reported to 5 decimal places in the spreadsheet. The second column is the magnitude of the object. This needs to be reported to 3 decimal places in the spreadsheet.

The third and fourth columns can be deleted. The fifth column is intended to give an idea of the quality of the data. Use the spreadsheet to sort this column and delete all the results that are not of quality A, B or C.

Save the spreadsheet, just the first two columns, in tab-delimited format.

THE GOLDEN RULE

There is a golden rule that anybody hoping to make a variable star discovery needs to remember and apply **before** starting detailed work on a light curve. You need to make certain that the star you are studying has not already been identified and published as being a variable star. Nobody will thank you for duplicating the work of other people.

There are two places where this can be done.

The first is at the International Variable Star Index as maintained by the American Association of Variable Star Observers (AAVSO). This can be found at:-

http://www.aavso.org/vsx/index.php?view=search.top

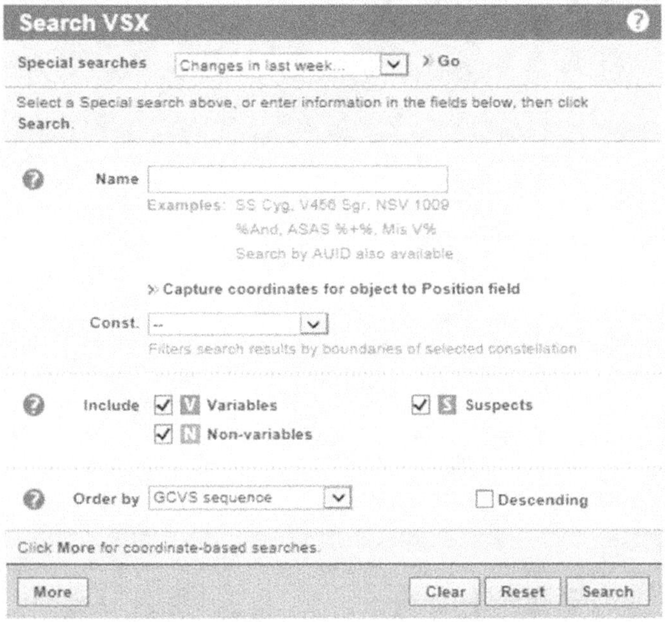

Fig. 4 - The VSX search facility

The first image shows the default option and the second the screen when a positional search is being done.

It is standard practice to do a co-ordinate or positional search with a search radius of 10 arc seconds

If you do a search and no matching star is listed you can be fairly certain that you are not wasting your time.

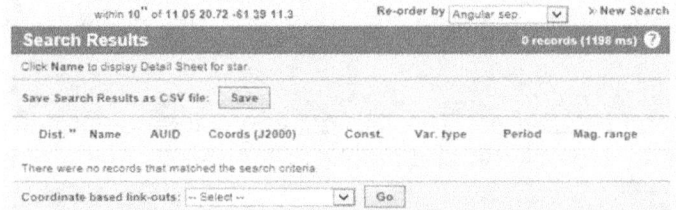

Fig. 5 – The VSX results page showing no matching star. A new discovery!

A useful alternative to the VSX site is offered by the Vizier site. This can be accessed here:-

http://vizier.u-strasbg.fr/viz-bin/VizieR-3?-source=B/vsx/vsx

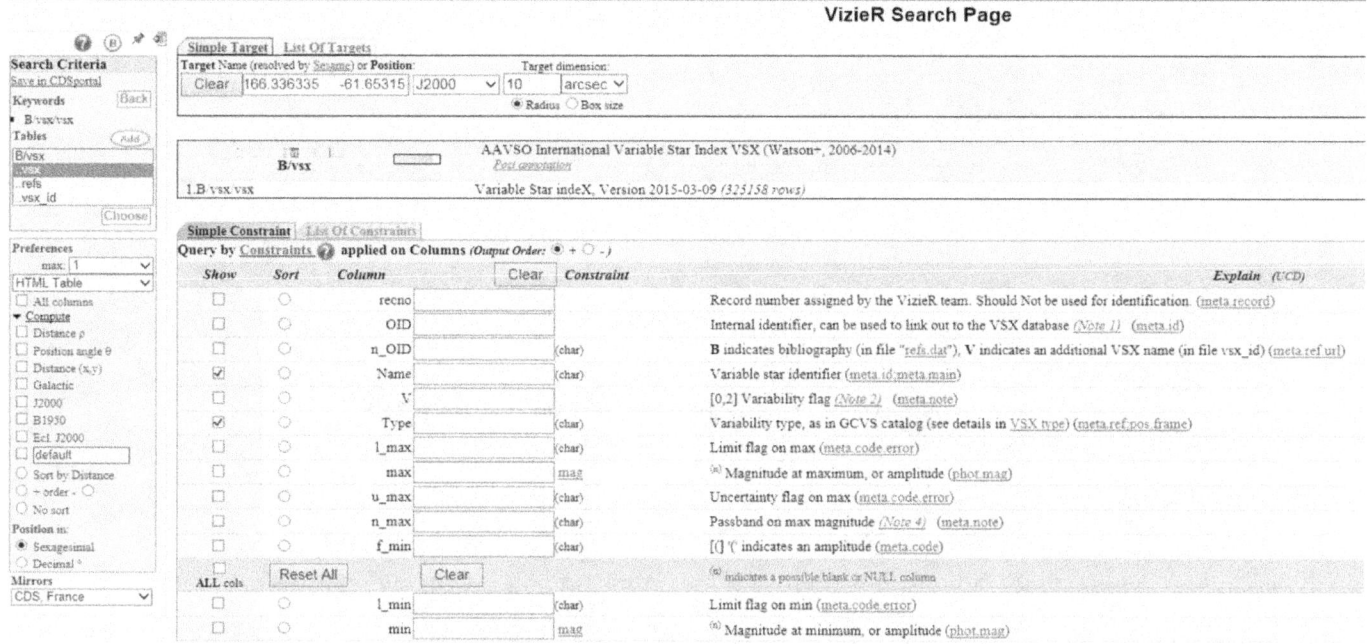

Fig. 6 – The Vizier input screen

This looks like a very complicated data entry screen. However all the researcher needs to do is to enter the positional data for the suspected variable star in the box at the top and then select Name and Type from the output options. If the star has already been reported the name and type will be reported, but a new discovery will report the words "No object found ….".

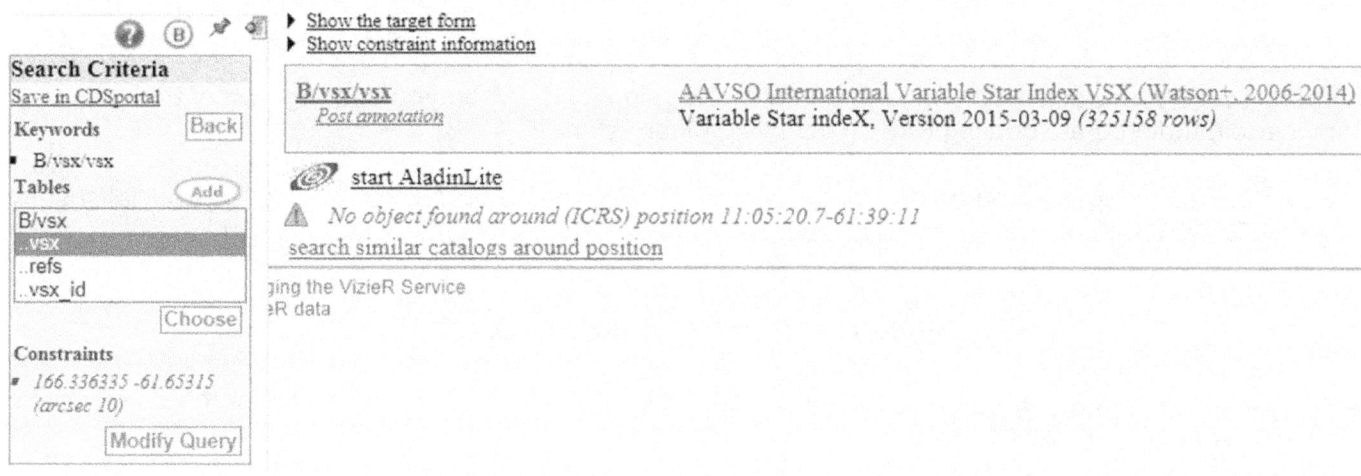

Fig.7 - The Vizier results page showing no matching star. A new discovery!

When you have satisfied yourself that your suspected variable star has not been reported already it is time to move onto more detailed analysis of the light curve.

A WORKED EXAMPLE

Field: CAR_SC3 StarID: 25880

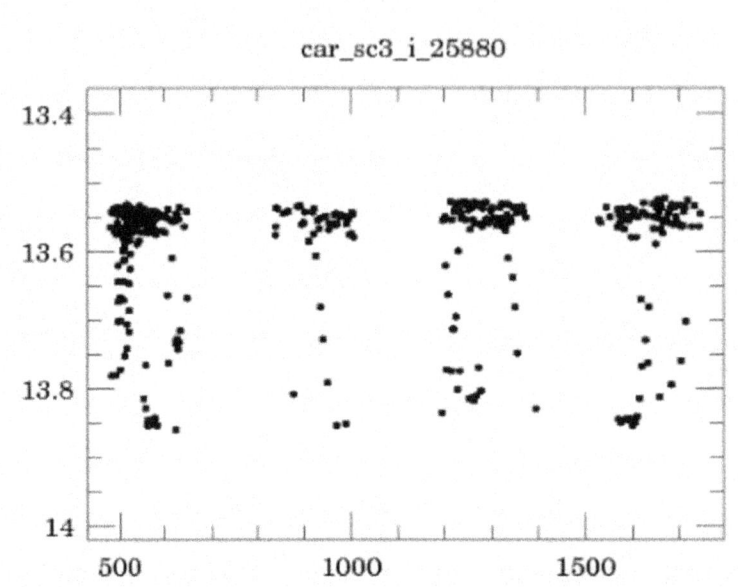

Fig. 8 – The light curve for VSX J110901.6-604830

VSX J110901.6-604830 is an eclipsing binary discovered by Jansen in 2010. Details of OGLE data relating to this star can be found at:-

http://ogledb.astrouw.edu.pl/~ogle/photdb/getobj.php?field=CAR_SC3&starid=25880&tmpdir=1230366828.79726s&db=DIA&points=good

The graph is a very characteristic shape that I call it the rain cloud. When you see a graph of this type you can be fairly certain that you are dealing with an eclipsing binary star.

Prepare a data file using the method described earlier. This file may be processed using the specialist software package PERANSO.

PERIOD ANALYSIS USING PERANSO

Peranso is a computer programme that can be used for the analysis of light curves and for the determination of the period of variability. It can be downloaded from:-

http://www.peranso.com/

The software comes with detailed instructions so I will not go into great detail within this worked example.

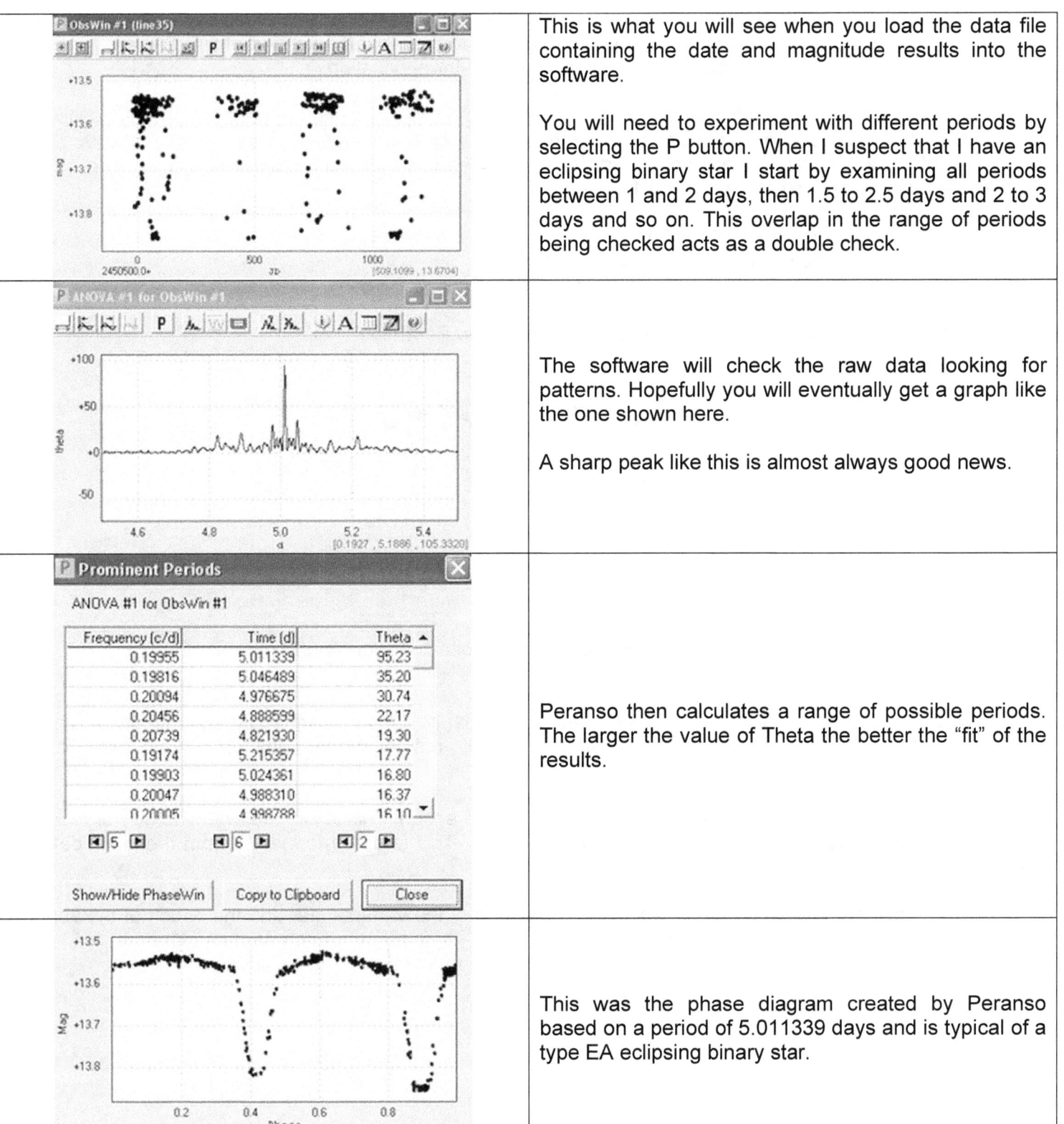

This is what you will see when you load the data file containing the date and magnitude results into the software.

You will need to experiment with different periods by selecting the P button. When I suspect that I have an eclipsing binary star I start by examining all periods between 1 and 2 days, then 1.5 to 2.5 days and 2 to 3 days and so on. This overlap in the range of periods being checked acts as a double check.

The software will check the raw data looking for patterns. Hopefully you will eventually get a graph like the one shown here.

A sharp peak like this is almost always good news.

Peranso then calculates a range of possible periods. The larger the value of Theta the better the "fit" of the results.

This was the phase diagram created by Peranso based on a period of 5.011339 days and is typical of a type EA eclipsing binary star.

Fig. 9 – Using the PERANSO software.

There is a standard format that should be used when presenting phase diagrams of this type. The mid-point of the primary minimum should be positioned at phase 0.0. This means that the version of the diagram generated by Peranso needs to be slightly tweaked before it is submitted for publication.

By right-clicking on the diagram it is possible to download the data points used to generate the plot. Suppose, for example, the first version of the phase diagram has a minimum brightness at a phase of 0.9. Simply subtract 0.9 from all the listed phase values and if this reduces any individual value below 0 just add 1 to the value to maintain all values between 0 and 1.

A phase value of 0.9 becomes a value of 0.0
A phase value of 0.95 becomes a value of 0.05
A phase value of 0.6 becomes a value of –0.3 and then a value of 0.7 when 1 is added to it.

All the other changes are cosmetic and are easily carried out in Peranso.

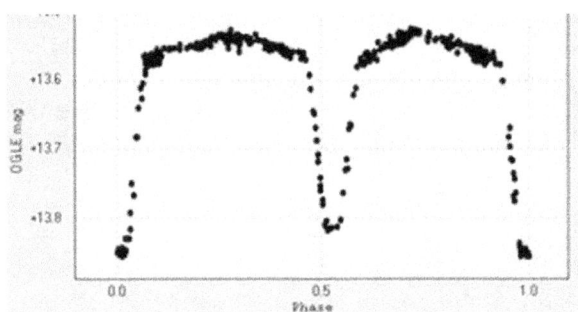

Fig. 10 – The finished diagram, suitable for publication

This is the type of diagram that needs to be generated. At a glance it is possible to see the magnitude range of the eclipsing binary (13.5 to 13.85) and to see that, unusually, the secondary minimum doesn't take place at phase 0.5.

The other key piece of data that needs to be determined is the epoch. This is just the scientific way of reporting a date when the variable star was at its minimum brightness. This is easily found out from the initial data file using the sort facility.

So you now know the position of the variable star, the period of the variable star and the epoch of the variable star. You are ready to publish your results!

PUBLISHING YOUR RESULTS

There are a number of alternatives the researcher might consider – each with some advantages and some disadvantages.

It certainly used to be possible to submit results directly to the International Variable Star Index. Helpful information on the submission process, that should be followed very carefully, can be found at:-

http://www.aavso.org/vsx/index.php?view=about.notice

"We encourage individual submission of new variables to the Variable Star Index (VSX). These can be single stars that you have discovered through your own observations or through data-mining; batches of stars, such as from a publication that is not already included in VSX; or changes/modifications/comments on existing VSX stars."

Providing the AAVSO mean what they say then even a single new variable star discovered through data mining can be submitted for inclusion in the Index.

Another option is the Open European Journal on Variable Stars. Again there is a website where information on the submission process can be found:-

http://var.astro.cz/oejv/

The OEJV expects rather more from the author/researcher but it is a valuable learning experience to write a "proper" scientific paper.

Only eight articles have been published in OEJV since January 2014 and I would strongly suggest that any prospective author makes contact with the editorial committee before starting work.

TYPES OF VARIABLE STAR

http://www.aavso.org/vsx/index.php?view=about.vartypes

WHAT HAPPENS AFTER YOU HAVE CONFIRMED YOUR DISCOVERY?

This is the only part of the process that gives major cause for concern. Once a new variable star discovery has been published in a peer reviewed journal the successful researcher is entitled to expect that it would quickly be included in the standard catalogue – sadly this is not the case.

Indeed the home page of the General Catalogue of Variable Stars makes fairly depressing reading –

http://www.sai.msu.su/gcvs/gcvs/index.htm

The single page list of "recent papers" on the GCVS site has papers that are up to twelve years old and there are peer reviewed papers, published by the AAVSO as long ago as 2009, that the staff responsible for the GCVS have yet to process!

EXAMPLES OF VARIABLE STARS FOUND IN THE OGLE ARCHIVES

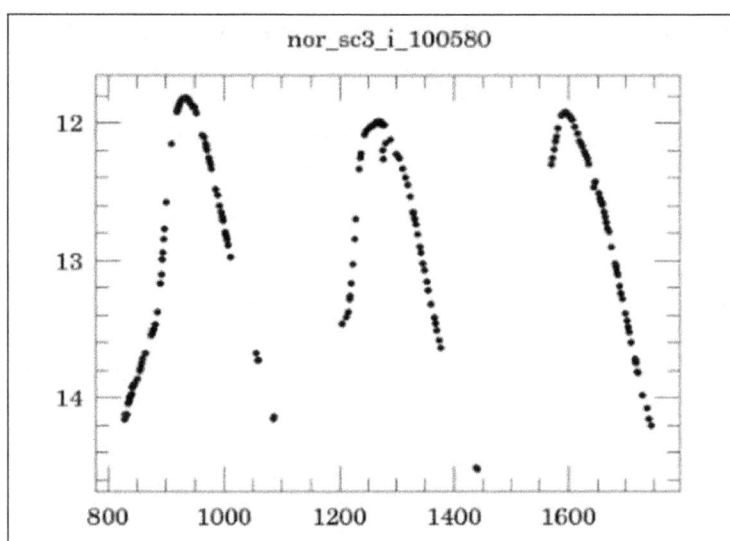

NOR_SC3 100580

Discoverer – M P Nicholson

Reference -
http://www.aavso.org/publications/ejaavso/ej91.pdf

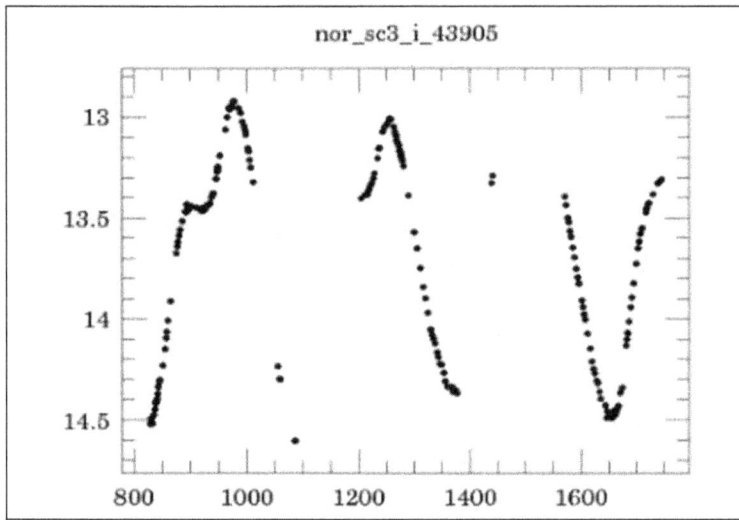

NOR_SC3 43905

Discoverer – M P Nicholson

Reference -
http://www.aavso.org/publications/ejaavso/ej91.pdf

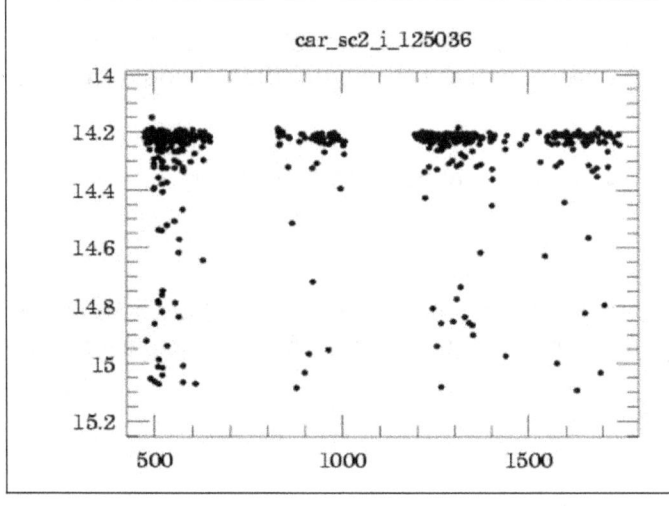

CAR-SC2 125036

Discoverer – M P Nicholson

Reference –
http://www.aavso.org/vsx/docs
/185549/194/ecl22curve.jpg

Fig. 11 – Peer-reviewed variable stars discovered in the OGLE archives

Carry out an OGLE search using the following constraints –

SELECT FROM disk2 WHERE Pgood>=95 and Pgood<=100 and I>=12 and I<=14.5 and Isig>=0.1 and Isig<=100

You should find that 402 objects have been listed. Work your way through the individual entries. How many of the stars are "clearly variable" and of these how many have already been listed in VSX?

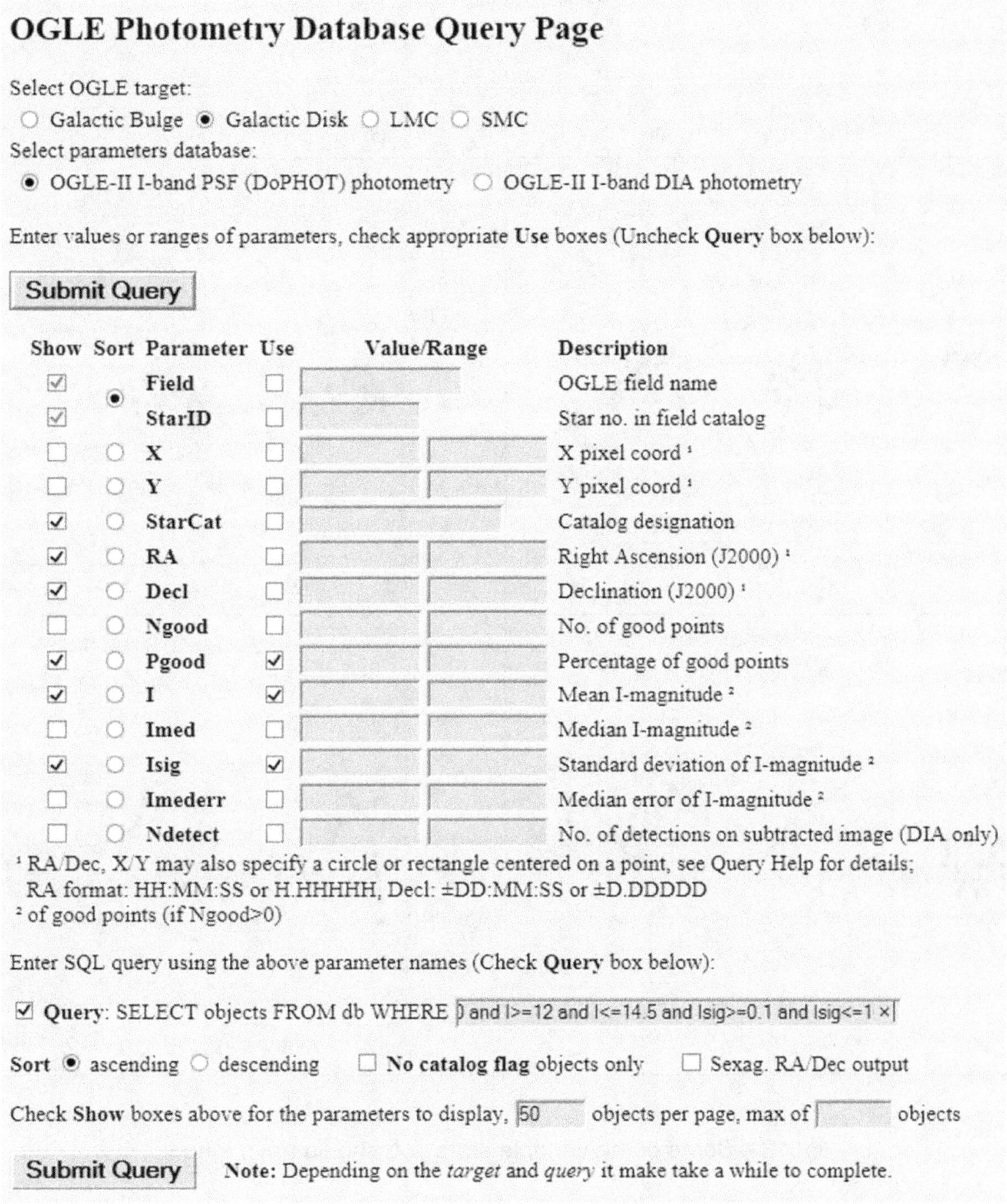

OGLE Photometry Database Query Page

Select OGLE target:
○ Galactic Bulge ● Galactic Disk ○ LMC ○ SMC
Select parameters database:
● OGLE-II I-band PSF (DoPHOT) photometry ○ OGLE-II I-band DIA photometry

Enter values or ranges of parameters, check appropriate **Use** boxes (Uncheck **Query** box below):

Submit Query

Show	Sort	Parameter	Use	Value/Range		Description
☑	●	Field	☐			OGLE field name
☑		StarID	☐			Star no. in field catalog
☐	○	X	☐			X pixel coord [1]
☐	○	Y	☐			Y pixel coord [1]
☑	○	StarCat	☐			Catalog designation
☑	○	RA	☐			Right Ascension (J2000) [1]
☑	○	Decl	☐			Declination (J2000) [1]
☐	○	Ngood	☐			No. of good points
☑	○	Pgood	☑			Percentage of good points
☑	○	I	☑			Mean I-magnitude [2]
☐	○	Imed	☐			Median I-magnitude [2]
☑	○	Isig	☑			Standard deviation of I-magnitude [2]
☐	○	Imederr	☐			Median error of I-magnitude [2]
☐	○	Ndetect	☐			No. of detections on subtracted image (DIA only)

[1] RA/Dec, X/Y may also specify a circle or rectangle centered on a point, see Query Help for details;
RA format: HH:MM:SS or H.HHHHH, Decl: ±DD:MM:SS or ±D.DDDDD
[2] of good points (if Ngood>0)

Enter SQL query using the above parameter names (Check **Query** box below):

☑ **Query**: SELECT objects FROM db WHERE) and I>=12 and I<=14.5 and Isig>=0.1 and Isig<=1 ×|

Sort ● ascending ○ descending   ☐ **No catalog flag** objects only   ☐ Sexag. RA/Dec output

Check **Show** boxes above for the parameters to display, 50 objects per page, max of ____ objects

Submit Query   Note: Depending on the *target* and *query* it make take a while to complete.

Fig. 12 – Advanced searching in the OGLE database

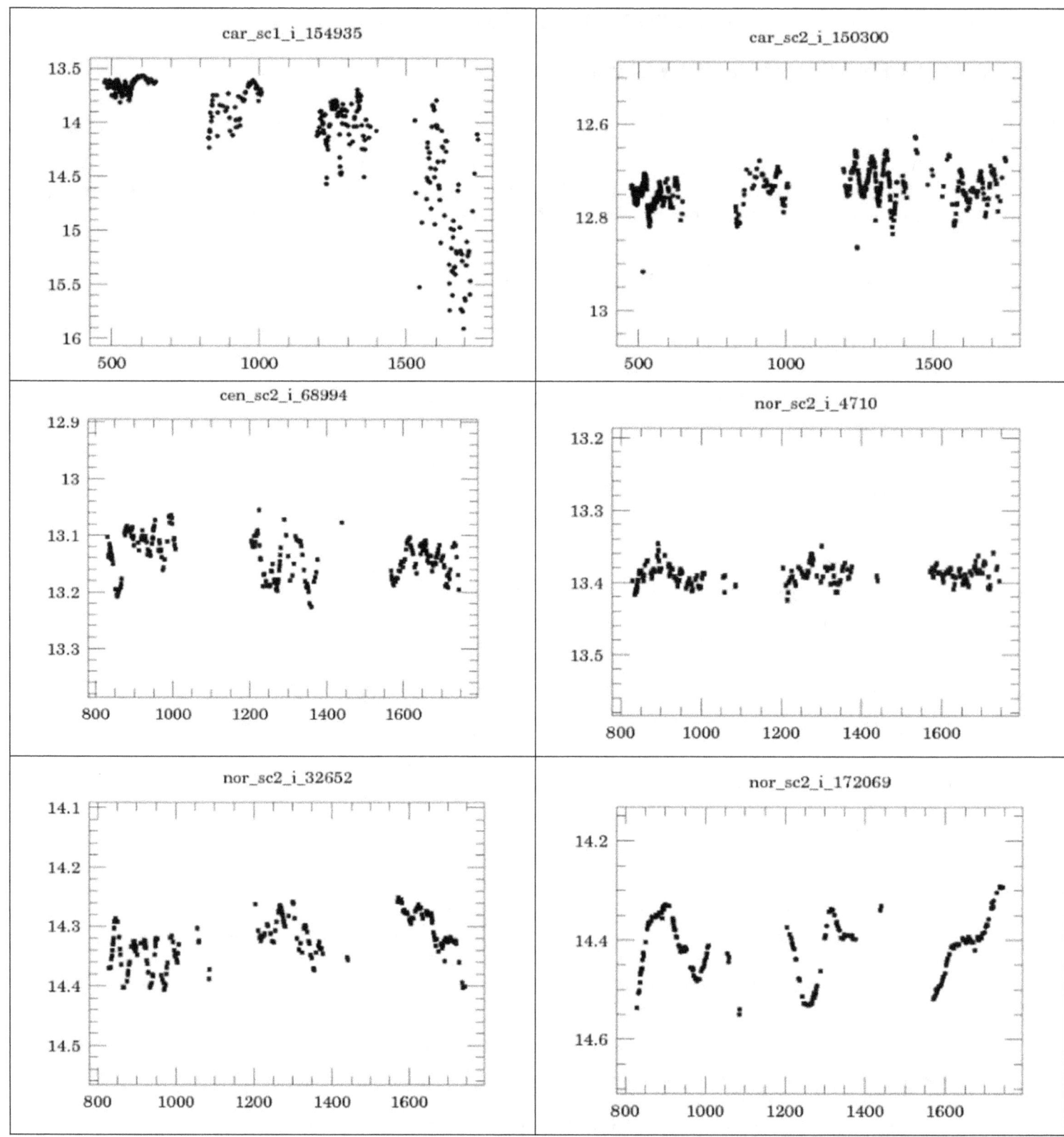

Fig. 13 – Some of the variable stars you should have found

What is the effect on the number of targets listed if you widen the scope of the survey to include fainter objects?

EXERCISE RESULTS

# Field	StarID	StarCat	RA	Decl	Pgood	I	Isig
NOR_SC1	232371	161333.59-535131.7	16.225996	-53.85882	95	12.004	0.118
NOR_SC3	221190	161636.97-541309.1	16.276936	-54.21919	95	12.005	0.204
NOR_SC3	178688	161607.76-534940.6	16.268823	-53.82794	95	12.011	0.25
NOR_SC3	221198	161636.27-541114.3	16.276743	-54.18731	95	12.02	0.234
NOR_SC3	58137	161525.29-533742.8	16.257025	-53.62855	95	12.067	0.242
NOR_SC4	221101	161749.70-540117.6	16.297139	-54.02156	96	12.084	0.101
NOR_SC3	43863	161530.89-534715.6	16.258581	-53.78765	95	12.25	0.288
NOR_SC3	264940	161632.50-533729.6	16.275695	-53.62488	95	12.26	0.122
NOR_SC3	16043	161522.26-540805.2	16.256183	-54.13478	95	12.263	0.112
NOR_SC4	12139	161654.11-540753.9	16.281698	-54.13164	95	12.269	0.195
NOR_SC3	16044	161522.48-540718.8	16.256246	-54.12189	96	12.302	0.347
NOR_SC3	221207	161628.42-541140.5	16.274561	-54.19459	96	12.307	0.104
NOR_SC4	16727	161644.36-540734.7	16.278988	-54.12629	95	12.315	0.31
NOR_SC1	223590	161332.52-535855.2	16.2257	-53.98199	95	12.318	0.15
NOR_SC3	264949	161625.90-533515.9	16.273862	-53.58775	98	12.344	0.171
NOR_SC4	7901	161651.06-541416.6	16.280851	-54.23795	96	12.375	0.168
NOR_SC3	43859	161519.88-534835.2	16.255522	-53.80978	97	12.411	0.16
NOR_SC3	20580	161522.84-540524.1	16.256344	-54.09002	96	12.438	0.162
NOR_SC3	205042	161605.82-533047.6	16.268283	-53.51323	96	12.448	0.22
NOR_SC3	161386	161620.42-540320.7	16.272338	-54.05574	95	12.48	0.112
NOR_SC3	225580	161635.66-540813.1	16.276572	-54.13698	97	12.48	0.109
NOR_SC4	225490	161801.51-540014.6	16.300421	-54.00406	97	12.48	0.118
NOR_SC1	240515	161317.87-534440.6	16.22163	-53.7446	98	12.487	0.213
NOR_SC3	238540	161629.05-535920.1	16.274736	-53.98892	97	12.527	0.175
NOR_SC1	159122	161301.41-535956.9	16.217059	-53.99913	95	12.539	0.177
NOR_SC5	4345	162244.61-524256.1	16.379059	-52.7156	95	12.555	0.103
NOR_SC3	251199	161638.13-534603.0	16.277259	-53.7675	95	12.605	0.106
NOR_SC4	233994	161808.13-535208.0	16.302258	-53.86889	95	12.605	0.103
NOR_SC6	182225	162442.64-520814.1	16.411843	-52.13725	95	12.644	0.102
NOR_SC3	3537	161519.89-541809.2	16.255525	-54.30255	96	12.647	0.166
NOR_SC3	242847	161638.02-535600.7	16.277228	-53.93352	95	12.647	0.111
NOR_SC1	40357	161214.78-535831.5	16.204104	-53.97542	96	12.677	0.112
NOR_SC3	260159	161631.98-534039.1	16.275551	-53.67753	97	12.684	0.298
NOR_SC3	10	161521.34-542200.8	16.255928	-54.36689	95	12.698	0.223
NOR_SC3	269540	161637.44-533344.2	16.277066	-53.56229	96	12.704	0.12
NOR_SC3	95906	161546.52-540111.1	16.262922	-54.01974	95	12.712	0.101
NOR_SC3	187104	161559.31-534335.2	16.266475	-53.72645	95	12.717	0.176
NOR_SC1	28940	161215.58-540949.8	16.204328	-54.16384	96	12.727	0.156
NOR_SC3	62891	161530.66-533204.1	16.258516	-53.53447	96	12.733	0.24
CAR_SC2	150300	110845.96-611318.4	11.1461	-61.22179	95	12.74	0.172

# Field	StarID	StarCat	RA	Decl	Pgood	I	Isig
NOR_SC4	12148	161648.48-540923.1	16.280134	-54.15641	95	12.769	0.209
NOR_SC3	43882	161515.98-534631.5	16.254439	-53.77541	96	12.77	0.158
NOR_SC3	95900	161554.02-540248.2	16.265004	-54.04672	95	12.775	0.22
NOR_SC3	82704	161546.93-541336.3	16.263037	-54.22674	95	12.777	0.102
NOR_SC4	251034	161802.20-533815.0	16.300611	-53.6375	96	12.783	0.143
NOR_SC3	39429	161520.22-535028.6	16.255617	-53.84128	95	12.784	0.12
NOR_SC3	182761	161622.33-534754.8	16.272869	-53.79855	97	12.794	0.232
NOR_SC4	53568	161650.56-533825.4	16.280711	-53.6404	95	12.801	0.119
NOR_SC4	161978	161745.66-540022.6	16.296017	-54.00629	96	12.844	0.106
NOR_SC3	170126	161600.85-535821.9	16.266903	-53.97276	95	12.85	0.18
NOR_SC4	21305	161648.58-540220.3	16.280162	-54.03899	96	12.858	0.11
NOR_SC3	34534	161520.08-535400.5	16.255578	-53.90013	96	12.869	0.122
NOR_SC3	95905	161534.93-540115.3	16.259702	-54.02091	95	12.878	0.133
NOR_SC4	181891	161726.54-534004.3	16.290705	-53.66787	95	12.878	0.175
NOR_SC4	96586	161718.94-535833.6	16.288593	-53.976	95	12.886	0.101
NOR_SC3	209728	161631.81-542319.8	16.275502	-54.38884	96	12.893	0.131
NOR_SC4	92214	161719.23-540053.2	16.288676	-54.01477	96	12.898	0.114
NOR_SC5	215946	162343.51-522021.6	16.39542	-52.33933	95	12.922	0.535
CAR_SC2	131767	110836.45-613620.4	11.143457	-61.60567	95	12.941	0.131
NOR_SC3	58143	161519.97-533749.7	16.255547	-53.63048	95	12.944	0.313
NOR_SC3	145231	161602.04-541815.0	16.267232	-54.30415	95	12.947	0.14
NOR_SC3	7	161517.97-542235.2	16.254993	-54.37645	96	12.95	0.132
NOR_SC3	34535	161522.83-535317.4	16.256342	-53.88816	97	12.95	0.125
NOR_SC4	254950	161754.01-533400.7	16.298337	-53.56687	95	12.959	0.164
NOR_SC3	82709	161556.26-541108.8	16.265627	-54.18579	97	12.964	0.116
NOR_SC3	152851	161605.31-541334.1	16.268142	-54.22615	97	12.968	0.14
NOR_SC4	251039	161804.78-533618.3	16.301327	-53.60507	97	12.974	0.139
NOR_SC3	16052	161529.60-540729.3	16.258224	-54.12482	95	12.986	0.251
NOR_SC3	20583	161531.21-540435.5	16.258668	-54.07653	96	12.986	0.129
NOR_SC4	34152	161647.92-535226.9	16.279977	-53.87415	95	12.988	0.2
NOR_SC4	234000	161753.26-535040.4	16.298127	-53.84454	95	12.996	0.1
NOR_SC3	39430	161518.98-535007.2	16.255273	-53.83534	96	13	0.25
NOR_SC2	217706	161455.45-541916.7	16.248736	-54.32132	95	13.005	0.605
NOR_SC4	212627	161805.66-540745.9	16.301572	-54.12942	96	13.011	0.25
NOR_SC3	165842	161622.44-540143.7	16.272899	-54.02881	95	13.013	0.134
NOR_SC3	213508	161637.55-541932.1	16.277097	-54.32558	99	13.014	0.271
NOR_SC3	260161	161636.35-534000.8	16.276763	-53.66689	96	13.02	0.126
NOR_SC4	221108	161807.17-540137.5	16.301992	-54.02708	97	13.032	0.115
NOR_SC4	229730	161800.43-535604.1	16.30012	-53.93447	97	13.032	0.142
NOR_SC2	217704	161457.59-541944.0	16.24933	-54.32888	95	13.041	0.584

# Field	StarID	StarCat	RA	Decl	Pgood	I	Isig
NOR_SC4	254953	161754.41-533349.0	16.298446	-53.56362	96	13.059	0.149
NOR_SC3	229957	161632.17-540512.1	16.275602	-54.08669	97	13.077	0.499
NOR_SC4	58597	161650.12-533244.0	16.280589	-53.54555	96	13.085	0.174
NOR_SC3	255858	161631.43-534343.3	16.275396	-53.72868	97	13.087	0.154
NOR_SC4	212620	161752.36-541012.5	16.297878	-54.17013	95	13.088	0.153
NOR_SC3	217290	161637.45-541534.8	16.277071	-54.25967	99	13.095	0.186
NOR_SC4	238332	161801.30-534654.4	16.300362	-53.78178	96	13.098	0.199
NOR_SC4	4050	161642.14-541703.7	16.278371	-54.28436	95	13.099	0.105
NOR_SC4	254951	161805.87-533356.9	16.301631	-53.56582	95	13.101	0.228
NOR_SC4	212619	161749.87-541027.2	16.297186	-54.17421	97	13.102	0.111
NOR_SC4	53571	161643.59-533659.6	16.278776	-53.61654	96	13.117	0.135
NOR_SC4	58590	161646.59-533401.0	16.27961	-53.56695	97	13.129	0.218
NOR_SC4	212626	161807.37-540754.8	16.302046	-54.1319	97	13.136	0.108
NOR_SC3	247087	161630.98-535044.0	16.275272	-53.84555	95	13.138	0.158
NOR_SC4	96601	161716.26-535928.8	16.287849	-53.99133	96	13.138	0.174
NOR_SC3	161375	161602.88-540548.3	16.267466	-54.09675	95	13.149	0.107
NOR_SC4	137695	161733.38-542010.0	16.292606	-54.3361	95	13.156	0.111
NOR_SC3	170120	161622.65-535926.4	16.272958	-53.99068	97	13.157	0.138
NOR_SC3	174516	161622.32-535328.4	16.272868	-53.89123	95	13.16	0.235
NOR_SC4	193287	161724.68-533110.3	16.290189	-53.51953	95	13.16	0.178
CEN_SC2	68994	140009.36-631315.1	14.002601	-63.22086	95	13.161	0.335
NOR_SC3	273972	161623.78-533114.4	16.273273	-53.52068	95	13.178	0.138
NOR_SC4	166018	161722.76-535534.3	16.289656	-53.92621	95	13.179	0.139
NOR_SC3	62890	161518.09-533208.2	16.255026	-53.5356	96	13.185	0.209
NOR_SC3	53430	161519.28-533936.3	16.255354	-53.66008	96	13.188	0.137
NOR_SC3	39427	161523.56-535127.0	16.256544	-53.85751	95	13.194	0.108
NOR_SC3	25326	161514.10-540144.7	16.253916	-54.02909	95	13.197	0.178
NOR_SC4	229742	161754.59-535547.7	16.298496	-53.92991	97	13.197	0.111
NOR_SC3	187100	161604.72-534447.1	16.267978	-53.74642	95	13.208	0.22
NOR_SC4	108618	161715.77-534906.5	16.287715	-53.81847	95	13.216	0.108
NOR_SC3	191365	161607.86-533840.2	16.268849	-53.64449	97	13.22	0.195
NOR_SC4	25791	161647.53-535831.9	16.27987	-53.97553	95	13.225	0.114
NOR_SC3	48713	161519.24-534207.4	16.255344	-53.70206	96	13.25	0.202
NOR_SC3	157118	161618.16-540758.2	16.27171	-54.13283	95	13.25	0.186
NOR_SC4	7936	161654.11-541204.9	16.281697	-54.20135	95	13.252	0.157
NOR_SC3	95925	161535.70-540135.3	16.259917	-54.02647	95	13.253	0.147
NOR_SC3	95951	161553.52-535943.2	16.264867	-53.99532	96	13.254	0.168
NOR_SC3	137190	161552.97-532820.6	16.264714	-53.47238	95	13.26	0.198
NOR_SC3	20591	161517.26-540504.9	16.254796	-54.0847	95	13.268	0.104

# Field	StarID	StarCat	RA	Decl	Pgood	I	Isig
NOR_SC4	247003	161804.25-534107.1	16.30118	-53.6853	97	13.278	0.215
NOR_SC4	193311	161736.41-532948.4	16.293448	-53.49679	95	13.279	0.113
NOR_SC2	46882	161350.07-535243.2	16.230574	-53.87867	96	13.298	0.122
NOR_SC3	43884	161532.79-534601.3	16.259107	-53.76703	95	13.301	0.163
NOR_SC4	166015	161729.79-535547.0	16.291609	-53.92971	95	13.303	0.159
NOR_SC4	104730	161718.27-535207.9	16.28841	-53.86886	97	13.331	0.102
NOR_SC4	21306	161642.24-540120.9	16.278401	-54.02248	95	13.355	0.18
NOR_SC3	238549	161626.96-535912.8	16.274154	-53.98688	95	13.366	0.119
NOR_SC4	193307	161728.31-533009.6	16.291197	-53.50266	96	13.374	0.102
NOR_SC4	25788	161650.07-535843.5	16.280574	-53.97876	96	13.375	0.126
NOR_SC2	32615	161345.25-540341.7	16.229237	-54.06158	95	13.378	0.144
NOR_SC4	53598	161648.57-533629.6	16.280159	-53.60822	96	13.383	0.157
NOR_SC4	53588	161655.70-533803.9	16.282139	-53.63441	95	13.387	0.188
NOR_SC4	53606	161642.78-533608.1	16.27855	-53.60224	96	13.401	0.125
NOR_SC2	4710	161347.75-542428.6	16.22993	-54.40793	95	13.403	0.177
NOR_SC2	4726	161348.69-542331.7	16.230191	-54.39213	95	13.403	0.164
NOR_SC4	193306	161731.29-533011.9	16.292024	-53.5033	95	13.407	0.194
NOR_SC4	238339	161801.59-534945.7	16.300441	-53.82936	95	13.411	0.104
NOR_SC4	212636	161801.02-541015.4	16.300284	-54.17095	98	13.428	0.144
NOR_SC3	221215	161631.23-541323.3	16.275342	-54.22313	97	13.429	0.133
NOR_SC3	100600	161555.70-535747.2	16.265471	-53.96312	98	13.445	0.13
NOR_SC3	213519	161624.46-541944.9	16.27346	-54.32913	97	13.445	0.191
NOR_SC3	113911	161551.56-534641.3	16.264322	-53.77814	98	13.46	0.202
NOR_SC3	213533	161630.73-541721.8	16.275202	-54.2894	98	13.46	0.13
NOR_SC4	58611	161653.25-533414.0	16.281457	-53.57056	96	13.461	0.121
NOR_SC3	200516	161617.39-533128.8	16.271498	-53.52467	97	13.463	0.177
NOR_SC3	91360	161537.74-540322.5	16.260482	-54.05624	96	13.481	0.158
NOR_SC3	118601	161553.60-534445.5	16.264888	-53.74598	95	13.488	0.134
NOR_SC3	182771	161619.41-534811.5	16.272059	-53.80319	98	13.494	0.122
NOR_SC3	217301	161628.32-541547.0	16.274533	-54.26307	96	13.496	0.178
NOR_SC4	21337	161655.69-540059.3	16.282137	-54.01647	95	13.496	0.322
NOR_SC3	3553	161513.73-541739.8	16.253813	-54.29439	95	13.502	0.2
NOR_SC3	20589	161515.57-540520.2	16.254325	-54.08894	96	13.508	0.506
NOR_SC4	216811	161754.55-540604.0	16.298487	-54.10111	98	13.51	0.137
NOR_SC4	212642	161750.68-540910.3	16.297411	-54.15286	96	13.514	0.129
NOR_SC4	204570	161759.03-541541.1	16.29973	-54.26143	99	13.553	0.179
NOR_SC4	181907	161735.84-534003.2	16.293289	-53.66755	95	13.571	0.142
NOR_SC4	173736	161729.29-534742.4	16.291468	-53.7951	95	13.577	0.186
NOR_SC4	34192	161650.59-535031.2	16.280719	-53.842	95	13.58	0.102
NOR_SC4	225500	161803.35-540030.1	16.30093	-54.00836	96	13.596	0.139

# Field	StarID	StarCat	RA	Decl	Pgood	I	Isig
NOR_SC4	193305	161726.38-533015.4	16.290661	-53.50428	95	13.597	0.224
NOR_SC4	92232	161709.64-540144.0	16.286011	-54.0289	95	13.602	0.155
SCO_SC3	65422	164308.25-445254.6	16.718959	-44.88183	98	13.607	0.598
NOR_SC2	46902	161347.60-535125.1	16.22989	-53.85697	95	13.619	0.13
NOR_SC4	141572	161731.15-541730.1	16.291986	-54.2917	96	13.627	0.103
NOR_SC4	242833	161804.43-534600.2	16.301231	-53.76673	95	13.628	0.155
NOR_SC3	269558	161634.62-533333.6	16.276284	-53.55934	96	13.642	0.122
NOR_SC4	166020	161728.66-535525.1	16.291296	-53.92364	96	13.66	0.11
NOR_SC4	129634	161701.46-532952.9	16.283738	-53.49803	95	13.663	0.224
NOR_SC4	141586	161728.63-541455.3	16.291286	-54.24869	95	13.665	0.167
NOR_SC4	108613	161719.79-534943.6	16.28883	-53.82877	96	13.667	0.11
SCO_SC4	49762	164406.07-442452.0	16.735021	-44.41444	96	13.673	0.144
NOR_SC1	202107	161337.40-541802.9	16.227054	-54.30081	96	13.679	0.145
NOR_SC4	48558	161650.15-533949.6	16.280598	-53.66377	98	13.679	0.117
NOR_SC4	193309	161724.04-532957.2	16.29001	-53.49923	95	13.685	0.136
NOR_SC4	258996	161750.63-533158.4	16.297396	-53.5329	95	13.7	0.101
NOR_SC4	58607	161644.44-533509.4	16.27901	-53.58594	96	13.704	0.14
NOR_SC4	229740	161747.18-535558.8	16.29644	-53.93301	96	13.706	0.104
SCO_SC3	50744	164246.53-442230.0	16.712924	-44.37501	95	13.707	0.304
NOR_SC4	12171	161640.57-540816.5	16.277937	-54.13793	95	13.71	0.116
NOR_SC4	63371	161644.57-533109.1	16.279048	-53.5192	96	13.71	0.117
NOR_SC4	92227	161721.41-540238.8	16.289281	-54.04411	96	13.721	0.126
NOR_SC4	254968	161753.64-533305.6	16.298232	-53.55156	95	13.725	0.239
NOR_SC4	238351	161758.77-534738.8	16.299658	-53.7941	96	13.731	0.209
NOR_SC4	149514	161746.08-541014.7	16.296133	-54.17075	97	13.739	0.108
SCO_SC3	95073	164327.53-445331.3	16.724314	-44.89203	95	13.751	0.612
NOR_SC2	4729	161343.03-542309.2	16.228621	-54.38588	95	13.758	0.135
NOR_SC4	68059	161657.61-532758.6	16.282668	-53.46628	97	13.761	0.158
NOR_SC4	68065	161657.45-532647.7	16.282626	-53.44659	95	13.77	0.12
NOR_SC4	12172	161645.06-540816.5	16.279184	-54.13792	96	13.772	0.125
NOR_SC4	145394	161726.85-541332.2	16.290792	-54.22562	95	13.773	0.137
NOR_SC8	58022	162710.25-515214.4	16.452848	-51.87066	95	13.774	0.595
NOR_SC4	141577	161746.05-541657.2	16.296125	-54.28256	98	13.776	0.129
NOR_SC2	59181	161347.26-534204.1	16.229794	-53.70115	97	13.778	0.415
NOR_SC4	68054	161646.58-532828.8	16.279605	-53.47468	98	13.783	0.269
NOR_SC4	221114	161804.20-540307.0	16.301167	-54.05196	97	13.796	0.122
NOR_SC4	149523	161744.13-540818.9	16.295591	-54.13859	95	13.807	0.111
NOR_SC4	53607	161649.30-533556.0	16.280362	-53.5989	97	13.808	0.297
NOR_SC4	225514	161801.95-535730.4	16.300541	-53.95844	97	13.845	0.169
NOR_SC4	25799	161641.74-535713.5	16.278261	-53.95374	95	13.846	0.102

# Field	StarID	StarCat	RA	Decl	Pgood	I	Isig
NOR_SC2	258168	161458.71-534756.1	16.249642	-53.79892	95	13.85	0.149
NOR_SC4	197171	161723.31-532751.8	16.28981	-53.46438	96	13.854	0.125
NOR_SC4	25797	161642.30-535737.7	16.278417	-53.96048	95	13.855	0.197
NOR_SC4	242854	161806.79-534306.5	16.301886	-53.71848	96	13.856	0.101
NOR_SC4	48542	161648.86-534225.9	16.280239	-53.70719	96	13.873	0.107
NOR_SC4	193320	161744.81-532859.8	16.29578	-53.48328	95	13.874	0.2
NOR_SC4	34166	161648.42-535331.8	16.280118	-53.89216	97	13.895	0.111
NOR_SC4	125297	161713.05-533347.5	16.286957	-53.56319	95	13.896	0.129
NOR_SC2	248699	161505.15-535510.6	16.251429	-53.9196	95	13.897	0.101
NOR_SC4	32	161646.31-541847.0	16.279532	-54.31304	98	13.9	0.109
SCO_SC3	65420	164308.15-445315.4	16.71893	-44.88762	97	13.903	0.669
NOR_SC4	166017	161723.77-535543.1	16.289936	-53.92865	97	13.905	0.125
NOR_SC4	251048	161807.51-533839.5	16.302086	-53.64429	95	13.908	0.123
NOR_SC4	21332	161651.43-540123.8	16.280953	-54.02328	98	13.911	0.121
NOR_SC4	25800	161649.35-535709.2	16.280375	-53.95257	97	13.915	0.121
NOR_SC4	247027	161806.01-533944.6	16.30167	-53.66239	98	13.915	0.155
NOR_SC4	63372	161651.22-533106.9	16.280896	-53.51857	97	13.917	0.114
NOR_SC4	129637	161707.06-532903.8	16.285294	-53.48439	95	13.918	0.128
NOR_SC4	25792	161649.53-535821.7	16.280426	-53.9727	96	13.922	0.115
CAR_SC1	154935	110649.77-613023.5	11.113826	-61.50652	95	13.923	0.472
NOR_SC4	185778	161729.01-533651.2	16.29139	-53.61422	96	13.93	0.111
NOR_SC4	247019	161804.85-534056.2	16.301347	-53.68227	98	13.934	0.154
NOR_SC4	254961	161756.54-533435.9	16.299038	-53.57664	97	13.943	0.141
NOR_SC4	238348	161751.83-534753.7	16.297731	-53.79824	96	13.946	0.123
NOR_SC4	234013	161753.31-535241.6	16.298141	-53.87823	96	13.95	0.225
NOR_SC2	267080	161459.33-534041.3	16.249814	-53.67813	95	13.952	0.373
NOR_SC4	58614	161658.14-533347.0	16.282817	-53.56305	97	13.952	0.101
NOR_SC4	79768	161714.77-541212.1	16.287437	-54.20335	95	13.956	0.14
NOR_SC4	34177	161650.15-535247.0	16.280597	-53.87971	96	13.962	0.232
NOR_SC2	172021	161423.22-540325.2	16.239784	-54.057	96	13.969	0.12
NOR_SC4	133936	161704.49-532826.2	16.28458	-53.47394	95	13.976	0.145
NOR_SC3	95941	161553.88-540031.5	16.264967	-54.00876	96	13.994	0.105
NOR_SC4	254958	161751.20-533450.5	16.297557	-53.58069	97	13.994	0.173
NOR_SC4	251052	161803.41-533611.2	16.300948	-53.6031	98	13.995	0.118
SCO_SC3	8453	164233.09-444136.2	16.709192	-44.69338	97	13.995	0.177
NOR_SC4	157924	161735.81-540138.7	16.29328	-54.02742	98	13.996	0.126
NOR_SC2	197161	161444.89-534119.1	16.245803	-53.68864	95	14.001	0.466
NOR_SC4	169919	161732.70-535110.7	16.292417	-53.85298	95	14.001	0.102
NOR_SC4	247021	161750.61-534027.2	16.297393	-53.67424	96	14.009	0.122
NOR_SC3	213525	161628.23-541848.9	16.274508	-54.31359	95	14.018	0.171

# Field	StarID	StarCat	RA	Decl	Pgood	I	Isig
NOR_SC2	192840	161424.45-534542.8	16.240124	-53.7619	95	14.02	0.497
NOR_SC4	63380	161647.66-532852.1	16.279906	-53.48113	97	14.023	0.132
NOR_SC2	271643	161503.49-533749.6	16.250968	-53.63044	95	14.028	0.128
NOR_SC4	181896	161729.40-534123.6	16.2915	-53.68988	95	14.029	0.171
NOR_SC4	258994	161756.35-533214.9	16.298986	-53.53748	95	14.032	0.102
CEN_SC2	6857	135837.10-631009.2	13.976972	-63.16923	95	14.036	0.327
NOR_SC4	166065	161730.10-535512.2	16.291694	-53.92006	96	14.041	0.143
NOR_SC4	92239	161706.08-540046.8	16.285021	-54.01301	95	14.045	0.117
NOR_SC4	181909	161726.86-533933.7	16.290794	-53.65936	95	14.047	0.139
NOR_SC4	189195	161726.68-533551.6	16.290745	-53.59766	95	14.049	0.129
NOR_SC4	208634	161751.23-541126.1	16.297563	-54.19059	97	14.05	0.154
NOR_SC4	53592	161641.60-533739.0	16.278223	-53.62749	96	14.06	0.215
NOR_SC4	189202	161732.89-533405.9	16.292469	-53.5683	97	14.064	0.136
NOR_SC4	68058	161654.42-532817.8	16.281783	-53.47162	96	14.071	0.152
NOR_SC4	108656	161700.49-534715.5	16.283469	-53.78765	96	14.073	0.181
NOR_SC4	254957	161809.13-533452.8	16.302536	-53.58132	95	14.079	0.126
NOR_SC4	26	161644.94-541960.0	16.279149	-54.33332	96	14.08	0.152
NOR_SC4	4067	161652.34-541616.6	16.281204	-54.27127	96	14.082	0.118
NOR_SC4	30014	161647.74-535640.6	16.279927	-53.94462	98	14.084	0.145
NOR_SC4	238338	161805.62-534959.5	16.30156	-53.83321	97	14.084	0.1
NOR_SC4	204566	161807.89-541631.8	16.30219	-54.27549	98	14.088	0.137
NOR_SC4	166037	161727.92-535654.0	16.291089	-53.94835	97	14.09	0.112
NOR_SC2	9367	161346.93-542005.7	16.229704	-54.33491	95	14.098	0.499
NOR_SC4	79765	161708.37-541349.1	16.285658	-54.2303	95	14.099	0.132
SCO_SC1	4505	163957.97-451159.5	16.666102	-45.19986	96	14.104	0.571
NOR_SC4	58613	161652.84-533408.7	16.281344	-53.56907	98	14.105	0.112
NOR_SC4	216823	161800.24-540435.8	16.300067	-54.07662	97	14.113	0.122
NOR_SC4	63365	161652.87-533157.5	16.281353	-53.53264	98	14.114	0.131
NOR_SC4	212644	161805.03-540812.2	16.301397	-54.13672	99	14.117	0.111
NOR_SC4	100706	161712.68-535520.0	16.286855	-53.92223	95	14.122	0.1
NOR_SC4	238363	161748.97-534953.5	16.296936	-53.83154	97	14.136	0.102
NOR_SC4	87895	161720.80-540711.3	16.289112	-54.11981	95	14.14	0.118
NOR_SC4	157968	161731.49-540056.7	16.292081	-54.01575	97	14.14	0.208
NOR_SC4	12170	161657.51-540823.8	16.282641	-54.13995	95	14.141	0.114
NOR_SC4	166071	161725.65-535441.9	16.290459	-53.91164	97	14.146	0.112
NOR_SC4	25822	161647.81-535758.4	16.279946	-53.96623	97	14.147	0.106
NOR_SC4	121282	161705.57-533658.6	16.284881	-53.61628	95	14.149	0.199
NOR_SC4	7930	161651.63-541255.3	16.281008	-54.21536	97	14.152	0.102
NOR_SC4	145440	161728.34-541115.8	16.291206	-54.18771	95	14.155	0.145
NOR_SC4	254960	161807.56-533436.7	16.302101	-53.57685	96	14.159	0.153

# Field	StarID	StarCat	RA	Decl	Pgood	I	Isig
NOR_SC4	92283	161709.62-540130.4	16.286005	-54.02512	96	14.161	0.107
NOR_SC8	170359	162738.38-515124.6	16.460662	-51.85682	96	14.162	0.585
NOR_SC3	109654	161554.64-535014.8	16.265179	-53.83745	96	14.164	0.107
NOR_SC4	238391	161807.60-534735.9	16.302111	-53.79331	95	14.164	0.134
NOR_SC4	259045	161754.30-533121.1	16.298416	-53.52253	95	14.167	0.222
NOR_SC4	251063	161749.61-533815.8	16.297114	-53.63772	97	14.168	0.11
NOR_SC4	100683	161714.54-535700.8	16.287373	-53.95022	97	14.177	0.145
NOR_SC4	125296	161701.12-533406.2	16.283643	-53.5684	95	14.183	0.125
NOR_SC4	251043	161803.35-533909.2	16.300931	-53.65255	97	14.184	0.137
NOR_SC4	254970	161751.97-533258.9	16.297769	-53.5497	95	14.185	0.199
NOR_SC3	62977	161520.39-533122.7	16.255665	-53.52296	95	14.189	0.101
SCO_SC3	65417	164301.62-445352.9	16.717116	-44.89802	95	14.195	0.707
NOR_SC4	133985	161716.03-532616.0	16.287786	-53.43778	95	14.197	0.154
NOR_SC3	137229	161556.79-532900.3	16.265775	-53.48342	95	14.199	0.107
NOR_SC1	232434	161321.80-535225.3	16.222723	-53.8737	95	14.201	0.127
NOR_SC2	120625	161421.54-534759.1	16.239315	-53.79975	95	14.203	0.106
NOR_SC4	247020	161808.23-534045.3	16.302285	-53.67924	95	14.204	0.163
NOR_SC2	176203	161434.16-535915.2	16.242823	-53.98757	96	14.205	0.153
NOR_SC2	107194	161414.40-540016.3	16.237334	-54.00451	95	14.206	0.149
NOR_SC4	48552	161655.72-534040.3	16.282144	-53.67787	97	14.207	0.157
SCO_SC3	95082	164328.29-445035.7	16.724524	-44.84325	96	14.207	0.684
NOR_SC3	48773	161516.19-534314.9	16.254496	-53.7208	95	14.216	0.115
NOR_SC4	100716	161702.30-535406.7	16.283972	-53.90186	95	14.219	0.152
NOR_SC4	173753	161736.67-534900.5	16.293519	-53.81682	95	14.224	0.101
NOR_SC4	30015	161657.91-535640.0	16.282752	-53.94445	96	14.225	0.1
NOR_SC3	225628	161635.56-540809.4	16.276545	-54.13596	95	14.228	0.119
NOR_SC4	204572	161808.69-541520.7	16.302413	-54.25574	97	14.23	0.1
NOR_SC4	234031	161805.05-535326.1	16.301404	-53.89058	95	14.231	0.303
NOR_SC4	247043	161751.95-534155.5	16.297764	-53.69876	95	14.231	0.138
NOR_SC2	244097	161459.60-540031.4	16.249889	-54.00871	95	14.233	0.129
NOR_SC1	40386	161219.13-535929.0	16.205313	-53.99138	96	14.241	0.107
NOR_SC4	92281	161713.79-540137.0	16.287164	-54.02694	97	14.242	0.141
NOR_SC4	212657	161809.74-540956.2	16.302704	-54.16561	95	14.245	0.133
NOR_SC4	229759	161759.28-535560.0	16.299801	-53.93333	97	14.246	0.125
NOR_SC4	242859	161749.82-534610.9	16.297173	-53.7697	96	14.246	0.189
NOR_SC4	157934	161734.98-540358.9	16.293051	-54.06636	97	14.247	0.104
NOR_SC4	25833	161647.93-535711.5	16.279982	-53.9532	97	14.248	0.182
NOR_SC4	129665	161719.88-533027.0	16.288856	-53.50749	97	14.254	0.164
NOR_SC2	217727	161457.81-541915.7	16.249392	-54.32102	95	14.257	0.443
NOR_SC2	55428	161344.25-534555.7	16.228958	-53.76548	95	14.263	0.168

# Field	StarID	StarCat	RA	Decl	Pgood	I	Isig
NOR_SC4	259038	161807.27-533145.2	16.302019	-53.52922	95	14.264	0.141
NOR_SC3	165885	161603.08-540125.6	16.267521	-54.02377	95	14.271	0.138
NOR_SC4	92286	161707.88-540115.0	16.285524	-54.02084	96	14.274	0.106
NOR_SC4	34223	161655.46-535116.7	16.282073	-53.85463	96	14.275	0.108
NOR_SC4	53575	161656.20-533922.3	16.282278	-53.65618	95	14.282	0.116
NOR_SC4	242909	161757.43-534328.2	16.299287	-53.72449	95	14.283	0.105
NOR_SC1	28983	161215.32-540940.9	16.204255	-54.16136	95	14.293	0.153
NOR_SC4	166079	161731.01-535400.7	16.291947	-53.90019	96	14.294	0.152
NOR_SC4	38136	161651.33-534658.6	16.280925	-53.78295	95	14.297	0.111
NOR_SC2	235161	161453.28-540719.9	16.248132	-54.12218	96	14.299	0.485
NOR_SC4	212662	161749.53-540935.6	16.297091	-54.15989	96	14.301	0.115
NOR_SC4	145408	161736.65-541340.1	16.293515	-54.2278	95	14.304	0.127
NOR_SC3	213546	161632.23-541856.7	16.275619	-54.31575	96	14.306	0.189
NOR_SC4	169958	161725.17-535106.5	16.290324	-53.85181	96	14.306	0.151
NOR_SC4	53621	161656.93-533831.8	16.282479	-53.64217	95	14.307	0.174
NOR_SC4	242911	161804.39-534325.2	16.301219	-53.72368	96	14.31	0.104
SCO_SC4	126452	164451.45-441423.4	16.747625	-44.23984	97	14.318	0.128
NOR_SC2	27654	161346.42-540723.2	16.229562	-54.1231	95	14.319	0.127
SCO_SC3	2319	164224.93-445345.7	16.706925	-44.89603	95	14.321	0.719
NOR_SC4	96635	161700.42-535816.5	16.28345	-53.97124	96	14.334	0.125
NOR_SC2	32652	161346.44-540344.0	16.229566	-54.06224	95	14.336	0.192
NOR_SC4	121280	161722.61-533717.2	16.289614	-53.62144	96	14.34	0.219
NOR_SC2	235163	161502.30-540714.2	16.250639	-54.1206	97	14.341	0.538
NOR_SC4	48588	161654.27-534053.1	16.281741	-53.68142	97	14.343	0.173
NOR_SC4	44155	161643.98-534307.4	16.278883	-53.71873	96	14.344	0.107
NOR_SC4	255000	161748.14-533343.6	16.296705	-53.56212	96	14.347	0.203
NOR_SC4	48603	161646.27-533950.7	16.279521	-53.66409	98	14.351	0.122
NOR_SC4	225541	161750.36-535838.5	16.297321	-53.97737	98	14.353	0.173
NOR_SC4	185794	161742.25-533811.1	16.295069	-53.63642	96	14.363	0.12
NOR_SC3	200559	161621.04-533134.6	16.272512	-53.52628	95	14.365	0.144
NOR_SC4	121270	161713.65-533751.1	16.287125	-53.63086	95	14.366	0.135
NOR_SC4	53620	161658.59-533837.1	16.282942	-53.64364	97	14.371	0.103
NOR_SC4	234053	161752.50-535146.5	16.297915	-53.86293	95	14.377	0.106
NOR_SC2	59209	161353.13-534243.2	16.231426	-53.712	96	14.378	0.374
NOR_SC4	259052	161800.71-533059.4	16.300196	-53.51649	96	14.378	0.134
NOR_SC4	129656	161720.21-533058.0	16.288948	-53.5161	96	14.379	0.193
NOR_SC4	63397	161657.41-533044.8	16.282614	-53.51245	97	14.381	0.119
NOR_SC2	59222	161344.43-534141.5	16.229008	-53.69487	96	14.386	0.544
NOR_SC4	25811	161642.18-535913.8	16.278383	-53.98716	96	14.386	0.122
NOR_SC4	177849	161729.98-534318.5	16.291662	-53.72179	95	14.392	0.169

# Field	StarID	StarCat	RA	Decl	Pgood	I	Isig
NOR_SC4	12198	161642.23-540928.7	16.278398	-54.15798	97	14.397	0.115
NOR_SC1	47250	161221.02-535245.8	16.205839	-53.8794	95	14.4	0.157
NOR_SC4	16796	161650.71-540437.2	16.280753	-54.077	98	14.407	0.103
NOR_SC1	54318	161214.20-534527.6	16.203946	-53.75767	95	14.409	0.104
NOR_SC2	37274	161342.98-535916.8	16.228606	-53.988	95	14.41	0.191
NOR_SC3	62962	161525.71-533219.8	16.257143	-53.53882	95	14.41	0.12
NOR_SC4	68075	161648.59-532839.5	16.280164	-53.47765	95	14.411	0.101
NOR_SC2	27685	161352.20-540720.3	16.231166	-54.12231	95	14.412	0.606
CAR_SC3	90806	111002.57-611222.6	11.16738	-61.20627	95	14.414	0.114
NOR_SC4	4081	161646.17-541736.5	16.279491	-54.29348	96	14.416	0.13
NOR_SC4	92275	161720.46-540207.5	16.289018	-54.03543	97	14.424	0.114
NOR_SC2	32664	161346.58-540220.6	16.229606	-54.03907	95	14.426	0.127
NOR_SC1	43761	161222.54-535726.0	16.206262	-53.95721	96	14.427	0.468
NOR_SC4	216830	161747.24-540728.5	16.296454	-54.12458	96	14.431	0.105
NOR_SC4	208608	161808.43-541332.6	16.302343	-54.22572	98	14.435	0.1
NOR_SC4	262944	161753.03-532752.9	16.298064	-53.46469	95	14.435	0.124
NOR_SC4	189222	161731.53-533513.3	16.292093	-53.58704	96	14.436	0.135
SCO_SC3	103693	164331.67-443520.7	16.725463	-44.58908	96	14.438	0.481
NOR_SC2	172069	161435.30-540217.3	16.243138	-54.03814	96	14.44	0.215
NOR_SC4	251068	161808.70-533801.3	16.302416	-53.63369	97	14.446	0.162
NOR_SC4	108638	161716.48-534903.4	16.287912	-53.81762	97	14.447	0.209
NOR_SC4	92291	161703.89-540053.4	16.284413	-54.01483	95	14.449	0.163
NOR_SC4	157969	161742.58-540050.3	16.295161	-54.01396	96	14.449	0.109
CAR_SC3	126169	111037.67-611709.4	11.17713	-61.28593	96	14.453	0.134
NOR_SC4	68087	161654.61-532707.9	16.281836	-53.45219	96	14.454	0.134
NOR_SC4	121271	161718.77-533743.2	16.288548	-53.62868	95	14.456	0.131
NOR_SC4	7979	161643.62-541113.6	16.278784	-54.1871	97	14.457	0.108
NOR_SC4	44122	161657.67-534531.2	16.282687	-53.75867	95	14.464	0.101
NOR_SC2	192857	161426.21-534517.6	16.240614	-53.75489	95	14.471	0.105
NOR_SC4	247054	161809.59-534010.4	16.302665	-53.66955	95	14.474	0.123
NOR_SC3	34569	161523.39-535500.8	16.256497	-53.91689	95	14.483	0.119
NOR_SC4	133978	161710.78-532702.1	16.286327	-53.45058	95	14.483	0.157
NOR_SC4	153741	161731.01-540600.0	16.291948	-54.1	97	14.483	0.103
NOR_SC2	217759	161505.26-541937.4	16.251461	-54.32705	95	14.486	0.433
NOR_SC4	125317	161712.30-533424.8	16.286751	-53.57357	95	14.487	0.16
NOR_SC1	40393	161222.59-535908.2	16.206275	-53.98562	97	14.488	0.146
NOR_SC4	181946	161742.34-534015.2	16.295093	-53.6709	96	14.49	0.132
NOR_SC4	254979	161757.18-533533.4	16.299217	-53.59262	97	14.49	0.165
NOR_SC4	68090	161651.64-532647.9	16.281011	-53.44665	95	14.494	0.135
NOR_SC4	12220	161651.53-540759.7	16.280982	-54.13324	96	14.499	0.109

# Field	StarID	StarCat	RA	Decl	Pgood	I	Isig
NOR_SC4	204592	161751.40-541646.2	16.297612	-54.27951	99	14.499	0.104
NOR_SC4	216849	161758.50-540606.2	16.299585	-54.10173	99	14.5	0.115

BY THE SAME AUTHOR

All are available from Amazon.com and from Amazon.co.uk

1800 new double stars for amateur observers

3600 celestial asterisms for amateur astronomers

Discover your own double star

Identifying Common Proper Motion Binary Star Systems

Identifying Identical Twin Star Systems from the SDSS Data Release 10